GUIDANCE MONOGRAPH SERIES

SHELLEY C. STONE

BRUCE SHERTZER

Editors

GUIDANCE MONOGRAPH SERIES

The general purpose of Houghton Mifflin's Guidance Monograph Series is to provide high quality coverage of topics which are of abiding importance in contemporary counseling and guidance practice. In a rapidly expanding field of endeavor, change and innovation are inevitably present. A trend accompanying such growth is greater and greater specialization. Specialization results in an increased demand for materials which reflect current modifications in guidance practice while simultaneously treating the field in greater depth and detail than commonly found in textbooks and brief journal articles.

The list of eminent contributors to this series assures the reader expert treatment of the areas covered. The monographs are designed for consumers with varying familiarity to the counseling and guidance field. The editors believe that the series will be useful to experienced practitioners as well as beginning students. While these groups may use the monographs with somewhat different goals in mind, both will benefit from the treatment given to content areas.

The content areas treated have been selected because of specific criteria. Among them are timeliness, practicality, and persistency of the issues involved. Above all, the editors have attempted to select topics which are of major substantive concern to counseling and guidance personnel.

Shelley C. Stone

Bruce Shertzer

CAREER EDUCATION AND BIOLOGICAL SCIENCES

LORETTA DIXON BRADLEY
PURDUE UNIVERSITY

HOUGHTON MIFFLIN COMPANY • BOSTON
ATLANTA • DALLAS • GENEVA, ILL. • HOPEWELL, N.J. • PALO ALTO

ISBN: 0–395–200474

Library of Congress Catalog Card
Number: 74–11972

CONTENTS

EDITORS' INTRODUCTION

Set IX of the Guidance Monograph Series is designed to treat the topic of career education and is intended for the use of teachers and counselors. The recent emphasis in educational circles on career education stresses that teachers are to have extensive input into the career education of their students from the elementary level through the secondary level. They are expected to draw out and teach the career implications of various subject matter areas. The contents of this set intentionally stress career education principles, procedures, techniques, and materials of value to the classroom teacher in specific subject matter areas.

Dr. Bradley's monograph presents a wide range of material relevant to the career education focused on the biological sciences. Of particular usefulness to those teaching in the biological sciences or those counseling students who are exploring potential careers is the material on careers directly and indirectly related to the discipline of biology. Also, the Appendix will supply the reader with valuable additional resources for use in classroom career education units or in the school's comprehensive career education program.

SHELLEY C. STONE
BRUCE SHERTZER

1

Why Career Education?

Why do people work? Perhaps the most common assumption is that people work to earn a living. While this answer seems logical, it is not absolutely correct to assume that people work only to earn a living. In fact, researchers (Centers, 1949; Lyman, 1955; Super, 1957) have reported that economic motive alone is not reason enough to explain why people work. Certainly these findings have a message for educators as they incorporate career education into their educational programs.

Within the last few years, a variety of programs emphasizing career education have emerged (Boocock, 1967; Conyne and Cochran, 1973; Hansen, 1972; Hoppock and Novick, 1961; McGuire, 1973; Morrill and Forrest, 1970). While career education models have varied, they all share a common goal — helping the student select a career that is best suited to his abilities, needs, and interests. Thus, they have been built on the premise that finding the right career and obtaining the proper training for that career is one of the most important decisions that an individual makes.

In a recent publication career education is defined as

> . . . the total effort of public education and the community aimed at helping all individuals become familiar in the values of a work

> oriented society, to integrate these values into their personal value system and to implement these values into their lives in such a way that work becomes possible, meaningful, and satisfying to each individual [Hoyt, Evans, MacKin and Mangum, 1972].

Certainly this definition has many implications for career education in the public schools. If for no other reason, one can argue that education cannot meet its purposes if career education is omitted. As early as 1944 some of these purposes were clearly identified by the Educational Policies Commission in a document entitled *The Purposes of Education in American Democracy.* While this document did not specifically illustrate a career education model, aspects of career education were included. The authors listed forty-three purposes of education. While the descriptions are too long to reproduce in this monograph, the following major headings seem to summarize those purposes (Norris, Zeral, and Hatch, 1960):

1. The Objectives of Self-Realization
2. The Objectives of Human Relationships
3. The Objectives of Economic Efficiency
4. The Objectives of Civic Responsibility

As can be inferred from this early educational model, education is seen as a program designed to develop the total individual. Specifically, education is viewed as self-development. Given this definition of education, career education is viewed as an integral part of the total educational program.

After acknowledging the need for career education, it seems appropriate to ask, what are the objectives of career education? Although objectives may vary among individual program settings, some objectives common to most programs have been listed by Hansen (1972).

> Students who experience this program (career education program) will become more vocationally mature persons who
>
> 1. are aware of their own preferred life styles and work values and can specify skills, abilities, attitudes, roles, and values congruent with those preferences.
> 2. exercise some control over their own lives through conscious choices and planning and can choose from alternatives a specific goal that is achievable and satisfying.
> 3. are familiar with the occupational options available in this society (major families or clusters of occupations grouped in some meaningful way), including the opportunity, structure, entrance requirements, and potential psychological satisfactions.
> 4. know the educational paths to and financial requirements for preferred occupations and careers.

5. are familiar with the process of career decision making.
6. know the major resources available in the school and community and are able to identify, use, and expand those resources (printed, human, and non-human) most appropriate for their particular goals.
7. can organize and synthesize knowledge about themselves and the work world and can map strategies for achieving their goals and modifying them if desirable.

Although Chapter 1 of this monograph attempts to identify the objectives of and the need for career education generally, the remaining chapters specifically relate career education to the biological sciences. Chapter 2 gives a brief history of the biological sciences while Chapters 3 and 4 describe careers directly and indirectly related to the biological sciences, respectively. The final chapter (Chapter 5) discusses the implementation of the career education program.

2

A History of the Biological Sciences

Every day millions of students are expected to make career choices. As suggested in the previous chapter, students need much information and assistance before they can make realistic career decisions. Students are more likely to have a better understanding of the various careers in the biological sciences if they have some knowledge of their historical development.

Realizing that there is a wealth of historical material available and that historical development should not be the major focus of a career unit, the author has attempted to present a brief summary of the historical development of the biological sciences. In any unit — even the briefest — different students have different interests. Since historical development will be of interest to some and of utility for others, it seems logical for the teacher to present some material on it. A reading list should be provided for students who desire a more complete coverage. Thus the presentation should help the students become more aware of the various careers and their relation to the development of the biological sciences.

The historical summary presented in this chapter is similar to the format and content presented in *A Short History of Biology* (Asimov, 1964). Basically the biological sciences are distinguished from other sciences because of their focus on the study of living things. Given this definition, it is easy to ascertain that early man had appreciation and respect for biology. Evidence found in embalmed mummies and art works certainly suggest that early man was interested in the plants and animals within his environment. In its early beginnings biology was constantly plagued by magic and superstition. Given the widespread use of the magical powers of the medicine man in curing disease, one can only conclude that he was prevalent in early societies and was apparently perceived as successful in his attack on disease. No doubt his concoctions were made from local plants, thus again suggesting an early but limited knowledge of biology.

Although much of what was known of early biology was not written, scientists know that the written code of Hammurabi existed as early as 1920 B.C. Basically this code described the rules and regulations for the medical profession. Another early contribution to biology, also associated with the medical profession, was the *Therapeutic Papyrus of Thebes* probably written about 1550 B.C. This book was written in hieroglyphics on papyrus above twelve feet wide and sixty-five feet long. Some of the topics discussed were diseases and treatments of the eye and ear, tumors, diseased organs, baldness, burns, diseases of children, and diseases of women. In addition, this document contained cosmetic prescriptions.

Classical Age

One of the most famous names associated with early biology is that of Hippocrates (460–377 B.C.). His abilities in using the medical art successfully have earned him the title of "father of medicine." Hippocrates believed that as a physician his task was to help nature to help the patient. Tradition had dictated prayer and sacrifice for healing, but Hippocrates stressed good diet, fresh air, fresh water, cleanliness, and exercise. Although he and his followers were not irreligious, they advocated the separation of religion and medical practice. Hippocrates and his students wrote more than fifty books. Probably the most famous is his book *On the Sacred Disease,* in which Hippocrates presented a rational approach to the cause and treatment of epilepsy, and thus denounced the age-old idea of a sacred cause. Many of his ideas have been incorporated into modern medicine. The oath taken by Greek physicians is still administered to physicians today.

The study of biology by the Greeks peaked with Aristotle (384–322 B.C.). Since Aristotle's father was physician to the ruling

Macedonian king, Aristotle had access to the best educational facilities at the court. Later he had access to the ideas of the Academy at Athens. Aristotle wrote more than one hundred forty books on many subjects, including politics, literature, physics, and biology. Many believe that the latter were his best writings, especially those relating to marine biology. Aristotle studied marine life and carefully described the habits, appearance, and development of sea creatures. From his many observations he arrived at his classification system or "ladder of life." Basically he grouped animals in a hierarchy with man standing at the zenith. Aristotle's classification was well ahead of its time and many of his concepts, such as dividing the scaly fish into two groups and classifying the dolphin with the land creatures, are maintained in today's modern classification systems. Thus it is no wonder that Aristotle has been acclaimed as the founder of zoology. While his interests did not lead him to study botany extensively, his student Theophrastus founded this science. Like Aristotle, Theophrastus carefully and accurately noted the appearances of organisms and classified more than five hundred plants.

Greek science flourished for four centuries and much progress was made. It was in the Alexandrian school of biology that the first human dissections were performed. The brain as the seat of intelligence, motor and sensory nerve differences, functioning of the retina of the eye, ovaries, liver and spleen were first studied there. While Greek science flourished for four hundred years, its decline began with the death of Alexander the Great (323 B.C.), and was complete by about 146 B.C. At this time the Greeks fell to Macedonian and later to Roman rule.

Although their conquests brought opportunities to study many plants and animals, biology did not seem to be the major concern of the Romans. Instead, attention was directed toward conquering and building an empire. Law, politics, and military activities were enhanced. As a result of these interests, Rome rewarded soldiers rather than scientists. Of course biology was not entirely ignored, but its major emphasis was limited to agriculture, medicine, and hygienic practices.

By far the most famous biologist of the Roman Empire was the Greek physician Galen, who lived from 130–200 A.D. Galen acquired most of his knowledge of anatomy while he was a surgeon at the gladiatorial arena. At the time Galen's beliefs were being made known, Christianity was establishing itself in the Roman Empire. Because of this religious influence, Galen was not permitted to dissect human bodies. Thus most of his anatomical knowledge was gained from the dissections of dogs, sheep, and apes. Galen believed that a "boiling process" heated the body and thus allowed for the circulation of blood. He thought blood was produced by the liver, and once the blood left the heart it never returned. Galen was such a revered physician that no one dared to question his work until several hundred years later when Harvey demonstrated circulation from and to the heart. Galen's book *On Anatomical Preparations* remained the standard medical text for 1200 years.

The Renaissance

During the Middle Ages biology was virtually dead, and in fact was not revived until the Renaissance. Foremost in this revival was the stress on human figures in art. To make depictions of the human figure seem real, artists had to be familiar with the anatomy of the body. Leonardo da Vinci (1452–1519) was probably the most famous of Renaissance artists. He was not only an excellent artist but also an excellent anatomist. He studied and made sketches of birds, insects, trees, flowers, and humans. Although most anatomists had followed the order established by Galen, Leonardo da Vinci established a new approach by dissecting fresh human bodies. He was thus able to study the body in more detail and represent it more accurately in his art.

The birth of modern biology and the "scientific revolution" occurred in the sixteenth century. Specifically, the year 1543 marks their beginning. It was in this year that Nicolaus Copernicus, a Polish astronomer, wrote a book in which he stated that the sun is the center of the solar system. This ended the old Greek view that established the earth as its center. A second work, *The Fabric (Structure) of the Human Body,* was published in 1554 by the leading anatomist of the Renaissance, Andreas Vesalius. Early in his training Vesalius questioned Galen's writings. When he learned that Galen had not actually dissected a human body, he was convinced that the Greek physician's writings were not completely accurate. Thus he began his own independent study of the human body. He made public dissections and demonstrated (by lecture during the dissection) the structure of the human body. His was the first accurate book of anatomy to be published. It contained not only descriptions but excellent illustrations which presented the brain, bones, and blood vessels in accurate detail. Vesalius met with great opposition for breaking with the Galen tradition. In fact, his views were labeled as heretical. In time, public pressure forced him to go to the Holy Land and he was killed in a shipwreck on his return.

Although Vesalius never realized the full extent of his accomplishment, he has been called the "father of anatomy." This title would probably have been given to Leonardo da Vinci if his works had not been lost until the eighteenth century.

Physiology

Although the Greeks had made a brief beginning toward the study of the functioning of the human body (physiology), many of their writings were incorrect. Specifically, their conclusions on the functioning of the heart were inaccurate. Their description of the heart as a pump proved to be accurate, but beyond this, especially concerning the passage of blood from the heart to the arteries and the veins, there were differing views. Italian anatomists suspected that Galen's suggestion that tiny holes permitted the passage of the blood through the heart was not correct, but before William Harvey's (1578–1657) discovery no one had suspected that valves delayed the backward flow of blood into the veins. Harvey performed his investigation by tying off an artery and observing that the side toward the heart bulged. On the other hand when he tied a vein, the side away from the heart bulged. Thus he concluded that the same blood was being pumped over and over again in one direction from the heart to the arteries to the veins and back to the heart to begin the circuit again. Harvey died before his theory on the circulation of the blood was really complete. Although he supposed that tiny vessels connected the arteries and veins, this was only a supposition because there was no way to see these tiny vessels. Capillaries could not be observed by the unaided eye. Their actual presence was reported by the Dutch scientist Jan Swammerdam and the Italian physiologist Marcello Malpighi. Their discovery was made by microscopic observation three years after Harvey's death. Although Swammerdam and Malpighi reported the first observation of capillaries, it was the Dutch merchant Van Leeuwenhoek who, while observing the tail of a tadpole under a microscope, saw the blood flowing from the arteries into the capillaries. Another important discovery, aided by the microscope, was reported by Robert Hooke (1635–1703). He observed that a thin slice of cork was made up of layers which he called cells. Although the use of the microscope offered unlimited possibilities, it was not used in any major discoveries for another hundred years. Its real effectiveness was impaired because it broke light into colors that prevented the viewing of fine details. It was not until 1820 that achromatic microscopes were invented.

Hooke's description of the cell was crystallized in 1838 when a German botanist, Matthais Jakob Schleiden (1804–81), stated that all plants are made of cells. While Schleiden was observing cells in various plants, a German physiologist Theodor Schwann (1810–82) was observing cells in animals. His conclusion, like Schleiden's, was that all animals are made of cells. Thus together Schleiden and Schwann formulated the cell theory.

Charles Darwin

Without a doubt, the book most influential to biology was *The Origin of Species*, published in 1859. This book was written by an English naturalist, Charles Darwin. After trying to study medicine and the ministry, Charles Darwin instead took a position as a naturalist aboard the ship "Beagle." During his five years aboard the "Beagle," he observed variation among species from one island to another. After his return he began to question the causes of the differences among species. For example, Darwin had observed a group of finches on the Galapagos Islands. While the finches were similar in many ways, Darwin noticed they were different in other ways, especially in their eating habits. One group ate seeds while another ate insects. Darwin observed 14 species in all. He could not accept the belief that this variation just occurred. After breeding pigeons to see the variation among the offspring, Darwin wrote his thesis. Basically he stated that the environment would place demands on the species for physical change (e.g. beak changes in the case of finches) until adaptation resulted in a new species that could survive within the environmental conditions. Thus the survivors were "fit" to live in the environment. By contrast, those species that could not meet the demands of the environment would die — "the survival of the fittest." At the same time that Darwin was writing *The Origin of Species*, another Englishman, Alfred Russell Wallace, was writing an almost identical theory. Wallace sent his manuscript for Darwin's consideration. Since neither of the men knew the other was even considering the topic, both were amazed at the similarity of ideas. Darwin's concept of evolution stimulated much protest. Many people thought it was in direct conflict with the story of creation as written in the *Bible*. Darwin avoided joining in the controversy. The crusade for the theory of evolution was done by Thomas Henry Huxley, and it was he who really took evolution to the common man.

Perhaps the greatest gap in Darwinian theory occurred with its omission of the hereditary mechanism. The solution to the hereditary problem was discovered by the Austrian monk Gregor Johann Mendel (1822–84). His experiments with garden peas were conducted in a small garden at the monastery. After crossing the plants grown from dwarf and tall seeds, he concluded that one trait (tallness) always seemed to dominate. Only when he bred dwarf with dwarf seeds did he obtain the dwarf pea plant. Mendel then went a step further and showed that in mating tall and dwarf plants, he could predict the number of tall, dwarf, and hybrid tall peas that would occur. Thus his genetic theory filled a void in the evolutionary theory. It illustrated that random variation did not average out over time, but kept appearing and reappearing in succeeding generations.

Medicine

Despite this involvement with evolutionary theory, man's energy was again diverted to medicine. Since early times man had been constantly plagued by disease, and in eighteenth century England the dreaded disease was smallpox. Smallpox spread over Europe and killed one out of three of its victims. It was from such an epidemic that Edward Jenner (1749–1823) discovered his vaccination for smallpox (1796). Although he found a method of vaccination, Jenner did not know what caused smallpox or, for that matter, any infectious disease. Louis Pasteur's "germ theory of disease" was not advanced for more than fifty years. Through the use of the microscope Pasteur was able to conclude that a disease is caused by microorganisms which spread from a sick to a healthy individual, who in turn becomes ill and causes the disease to spread to someone else. Pasteur stressed that disease microorganisms must be destroyed through hygienic procedures. Despite Pasteur's germ theory and its insistence on cleanliness, many people were still dying from infection after surgery. The solution to this problem was found by an English physician, Joseph Lister (1827–1912), who suggested that the sterilization of the surgical incision would prevent infection. Lister's technique was successful and marks the beginning of "antiseptic surgery."

Pasteur's germ theory of disease transmission could be applied to organisms other than bacteria which could cause a disease. Working with dissected mosquitoes, Ronald Ross was able to prove that the anopheles mosquito carried a parasite that produced malaria. A similar experiment was conducted by Walter Reed, who found that the aedes mosquito carried yellow fever.

Mental Illness

One area of biology that did not lend itself to the application of Pasteur's germ theory was that of mental illness. Mental illness had been known to exist for centuries, and in fact was discussed in writing by Hippocrates, yet this did not mean that it was under control. Rather the contrary seemed to prevail. Men viewed the mentally ill individual as one possessed by the devil, and cruel treatment rather than understanding was prescribed. The real beginning of a change in attitude toward mental illness occurred with Philippe Pinel (1745–1826), who classified it as a form of illness instead of possession by demons. A similar view was taken by an Austrian physician, Friedrick Anton Mesmer (1733–1851), who cured by a form of exorcism. Mesmer found that he could "free" his patient's mind of the illness by having him fix his attention on some object. This enabled the patient to concentrate on what the therapist was suggesting. In the beginning mesmerism, as it was called, was successful, but later, as Mesmer began to depend heavily on mysticism, he was forced to abandon his treatment of the mentally ill.

Despite Mesmer's unpopularity, his treatment was not lost. A Scottish surgeon, James Baird (1795–1860), renamed it hypnotism and the field of psychiatry entered the medical profession. Psychiatry really did not enjoy much prestige until it was adopted by an Austrian physician, Sigmund Freud (1856–1939). Freud used hypnotism to treat mental illness. He found that under the influence of hypnotism, the patient was freer to reveal the contents of his unconscious thinking. Although Freud enjoyed much success with hypnotic treatment, he found free association to be a more satisfactory approach. Free association aided the patient in revealing the inner contents of his mind, while at the same time making him consciously aware of his state. Freud felt that the understanding of his motives would allow the patient to correct maladaptive behavior. Thus Freud founded psychoanalysis, the analysis of the unconscious mind.

Another track for studying the mind was developed by the psychologist Wilheim Wundt. Although his work, like Freud's, was concerned with the workings of the mind, he was interested in studying behavior. Specifically he was trying to determine uniform descriptions of human behavior. His focus was on similarities in behavior, not differences. In conducting his research, Wundt measured sensitivity to visual and auditory stimuli and reaction time. Much of his research involved maze-running by rats. These results were then applied to man and later inspired Wundt's attempt to measure human intelligence.

Other studies relating intelligence and behavior to the nervous system were conducted by the Russian physiologist Ivan Petrovich Pavlov. In his early career he studied the secretion of digestive juice. No doubt this early interest in secretion led to his classic study of conditioning. Pavlov found that a hungry dog will salivate in the presence of food. Additionally, he found that if a bell is rung every time that food is presented, the dog will eventually associate the presence of food with the ringing of the bell. That is, the dog will salivate when the bell is rung regardless of whether food is or is not present.

Twentieth Century Advancements

The late nineteenth and beginning twentieth centuries saw spectacular accomplishments in the area of serology, the study of blood serum. During a diphtheria epidemic in 1892, a German bacteriologist, Emil Adolf von Behring (1854–1917), infected an animal with diphtheria. This caused the animal to produce antibodies. The antibody serum was then injected into a child infected with diphtheria. The treatment was successful. Previously diphtheria had killed almost all of its victims. A similar antigen antibody reaction allowed the German bacteriologist August Von Wasserman (1866–1925) to find a diagnostic device for syphilis. Another important serological discovery was made by the Austrian physician Karl Landsteiner (1868–1943). In 1902 Landsteiner discovered the four blood groups, which he named A, B, AB, and O. This paved the way for safe blood transfusions by making it possible to determine in advance whether a patient could or could not receive blood safely from a donor.

Another advancement by twentieth century serologists occurred with the discovery that disease can be induced by viruses. Viruses cause such diseases as polio, measles, mumps, chicken pox, influenza, and some types of cancer. Although cancer has become one of the greatest killers of man, its real cause has essentially remained unknown with the exception of an experiment conducted by an American physician, Francis Peyton Rous (1879–   ). Rous tested a chicken tumor to see if a virus were present. It was, and this experiment marked the first method for isolating an agent that could produce cancer. Despite this discovery, the cause and cure of cancer are still not clearly understood. In fact, it has been suggested that there are different types of cancer caused by different agents. Thus the discovery by Rous represents only a beginning, and not a conclusive understanding of the disease.

The recently developed penicillin proved invaluable in the discovery of vaccines for curing mumps and poliomyelitis. Cultures of the virus causing mumps could not grow because of the rapid growth of bacteria within the culture. To curb the growth of the bacteria, Enders (1897–) added some penicillin to the culture, and this stopped the growth of bacteria without altering the viral growth. One year later, in 1949, Enders used this technique on the poliomyelitis virus, and again it worked. But it was not until 1957 that the Polish-American microbiologist Albert Bruce Sabin (1906–) discovered a viral strain for each of the varieties of poliomyelitis. In the early 1960s Lawrence Katz (1927–) made a similar discovery concerning measles.

In the early 1930s techniques for understanding the chemistry of the cell and its internal structure were devised. Hans Adolf Krebs (1900–) worked out the famous "Krebs' cycle" which explains the conversion of lactic acid to carbon dioxide and water. At about this same time the first electron microscope was built. Since its operational principle is based on focusing electron beams rather than light beams, this new device afforded greater magnification than the previous microscopes.

Shortly after the invention of the electron microscope, biochemist Rudolf Schoenheimer (1898–1941) used radioactive isotopes to investigate metabolic reactions in animals. His early work led to such important discoveries as fat storage, stages of photosynthesis, and dating (fossils, etc.) by carbon 14.

As scientists probed more deeply into the chemical make-up of cells, such techniques as ultracentrifugation and electrophoresis were discovered. These techniques allowed for the separation of proteins. Thus scientists were able to separate out such fractions as albumin, alpha, beta, and gamma globulins. As might be expected, refined x-ray beams proved invaluable in the study of amino acid arrangement. In 1951 an American chemist, Linus Pauling (1901–), presented the amino acid arrangement and showed that its formation chain was in the shape of a helix. The mathematical computations for this data grew more complicated as scientists continued their investigations. The development of electronic computers in the 1950s was an invaluable asset.

The turning point in the study of molecules came in the early 1950s with the study of the DNA (deoxyribonucleic acid) molecule. Three scientists, Maurice Hugh Frederick Wilkins

(1916–), Francis Crick (1916–), and James Watson (1928–), were able to identify the components of the helical structure of the DNA molecule. This discovery of a double helix structure paved the way for the explanation of mitosis, the method by which chromosomes are doubled. Basically, each DNA molecule formed a replica of itself. When the DNA molecules did not carry out their replication process correctly, an imperfection or mutation occurred. With improved models of the electron microscope, scientists discovered that genetic information from the chromosomes reached the ribosomes through the "messenger" RNA (ribonucleic acid). Even with the wealth of knowledge obtained in the 1950s, the complete genetic code has not been broken. Today's scientists are still actively seeking the answer to this question and, for that matter, many others. Perhaps Isaac Asimov vividly summarizes the present situation when he writes:

> It is perhaps the most satisfying aspect of scientific work that no matter how great the advances or how startling and smashing the gains of knowledge over the unknown, what remains for the future is always greater, still more exciting, still more wonderful. What may not yet be revealed during the very lifetime of those now living? [Asimov, 1964]

After reviewing the material presented in this chapter, the teacher will need to decide whether to include a brief or a thorough coverage of the historical development of the biological sciences. Basically the question that faces the teacher is: How significant is historical development in the career decision-making process? Certainly it should be included within the career unit, but, in most instances, the greatest emphasis should be devoted to a discussion of career options, such as those discussed in Chapters 3 and 4. It is the author's opinion that the key objective is helping students to become more aware of and familiar with the various careers in the biological sciences, and this objective is the major focus of the next two chapters.

3

Careers Directly Related to the Biological Sciences

The term biological science refers to that branch of science that studies living organisms. Because of its emphasis on life, this area of science has often been called life science. Since the biological science field is vast, there are several specialities within it. Traditionally the specialities within the biological sciences were divided into two major divisions: botany, the study of plant life; and zoology, the study of animal life. Later divisions listed a third area: microbiology, the study of microscopic organisms. Recent classifications divide biology into more detailed divisions which emphasize occupational speciality.

The arrangement and description of occupational specialities in this monograph are similar to the format of the *Dictionary of Occupational Titles*. Each occupational listing includes a description of the occupation, worker tasks, work setting, and *Dictionary of Occupational Titles* code. Occupational descriptions are separated into two categories based upon whether or not the occupation requires college training. It should be acknowledged that in some instances the educational category into which the occupation should be placed is not clear. The final decision for placement of the occupation within an educational category was based upon the training that is most commonly required for that occupation. Thus the reader should be aware that while an occupation may be included within the grouping not requiring college training, some of the persons employed in that occupation may have had extensive college training. All occupations have a six-digit *Dictionary of Occupational Titles* (DOT) code. It is suggested that the reader who is unfamiliar with this code refer to Volume II of the DOT entitled "Occupational Group Arrangement of Titles and Codes." Additional assistance for interpreting the code may be obtained from a guidance counselor.

Botany

Botany is that branch of the biological science family which specializes in the study of the life functioning of plants. It is concerned with all aspects of plant life, including morphology, heredity, structure, classification, development, and physiology.

Occupational Descriptions: Occupations Requiring College Training

Agronomist *DOT Code 040.081*

Conducts research in field crop production. Is generally concerned with methods for obtaining better crop-yield in terms of both quality and quantity. May also conduct research on breeding (sometimes in specific geographic area) to understand not only the growing and breeding cycle but also the effects of climatic conditions on plant adaptability.

Biostatistician *DOT Code 020.188*

Conducts experiments to obtain statistical data for use in biological experiments. Studies the principles related to the life processes of plants. By using statistical techniques, the biostatistician compares the relationship between groups of organisms. Obtains statistical data relating to the cause and effect of a specific disease. Conducts research to determine the efficiency of a particular treatment plan. Biostatisticians may specialize in the study of plants, animals, or both plants and animals.

Botanist *DOT Code 041.081*

Studies the life processes of plants, which include their physiology, heredity, reproduction, growth, and development. May work outdoors in search of better growing conditions for plants or indoors studying cellular functioning. Work often has application to the fields of medicine, zoology, ecology, or agriculture.

Curator *DOT Code 102.118*
(Curator, Horticultural Museum; Curator, Herbarium)

Responsible for organizing collections of a museum. Maintains constant contact with scientists and organizations in an effort to exchange scientific information. Conducts research on new specimens acquired by the museum. May publish research findings. Organizes field trips. Is responsible for the administration of the museum and its employees.

Economic Botanist *DOT Code 041.081*

Conducts work similar to that of botanist with emphasis placed on the economic aspects of plants.

Forester *DOT Code 040.081*
(Forest Ranger)

Directs the management of forest land. Is responsible for mapping the forest area. Predicts future timber growth and makes recommendations for the planting of new trees. Manages the tree removal program to be sure that only specific trees are removed. Supervises fire prevention programs and directs the activities of workers during forest fires.

Geneticist *DOT Code 041.081*

Studies heredity. Specializes in the study of the mechanisms by which biological characteristics are transmitted from one generation to another. Performs experiments to determine the biological factors prominent in the origin, development, and transmission of inherited characteristics. Conducts research to improve plant breeds. Seeks methods for altering traits or producing new ones. May study both plants and animals or may specialize in the study of animals (Animal Geneticist or Animal Breeder) or of plants (Plant Geneticist or Plant Breeder).

Plant Ecologist *DOT Code 041.081*

Conducts research similar to that of botanist with emphasis on such environmental conditions as the effect of rainfall and temperature on plant growth.

Plant Nematologist *DOT Code 041.081*

Studies roundworms (nematodes). Emphasis is placed on the study of roundworms that produce plant disease. Classifies nematodes by structure, life existence, and location. Investigates nematode behavior and habitat within different soils to further isolate them and to develop control methods.

Plant Paleontologist *DOT Code 024.081*

Studies fossilized remains of plants. Assembles and classifies plants by their age, family, and evolutionary development. Investigates plants as seen in their formations in the earth's crust. Attempts to explain present, past, and future evolutionary developments.

Plant Pathologist *DOT Code 041.081*

Studies plant disease. Conducts research to determine the cause, control, and treatment of plant disease. Investigates the tendency of some plants to become diseased. Compares and contrasts the environmental conditions, physiological make-up, and treatment susceptibility between diseased and healthy plants. May specialize in the study of a specific plant or specific cause of disease.

Plant Physiologist *DOT Code 041.081*

Studies the functioning and biological activities of plants. Investigates such life processes of plants as metabolism, reproduction, and respiration. Studies nutritional (vitamins, minerals) and environmental (temperature, light, soil) conditions to ascertain their influence on the life functioning of plants. May specialize in the study of a specific plant.

Plant Taxonomist *DOT Code 041.081*

Identifies and classifies plants. Dissects plants to obtain more accurate information about their structure. Often specializes in the study of a specific plant. Often studies new species for classification.

Teacher, Botany *DOT Code 041.081*
(Public School or College)

Instructs students in the fundamental concepts of botany. May conduct research (especially at college level) relating to the life functioning of plants. May specialize in the study of one plant or in the study of a specific life process.

Occupational Descriptions: Occupations That Do Not Require College Training

Drug Grower *DOT Code 405.181*

Grows various plants (mint, digitalis) that are used by the medical and food industries. Purchases supplies and equipment needed for growing plants. Decides when crops will be planted and harvested. Maintains marketing records. Supervises field employees.

Floral Designer *DOT Code 142.081*
(Florist)

Selects flowers, foliage, and other accessories to be used in floral arrangements. Is responsible for designing floral arrangements (corsages, wreathes, bouquets). Directs the floral design for special events (funerals, weddings, etc.).

Forest-Fire Fighter *DOT Code 441.887*

Works to suppress forest fires. After fire has been extinguished checks to be sure that hot areas do not restart the fire. Fells trees and digs trenches. May work alone but usually works as member of fire-fighting crew.

Forester Aid *DOT Code 441.384*

Observes and records forest data such as tree growth, tree mortality, timber cutting, timber sales, timber damage by fire. Records data for water supply, rain level, and general soil features in timber areas. Assists in enforcing fire prevention regulations.

Grounds Keeper *DOT Code 407.884*

Maintains grounds on public, industrial, or commercial business property. Is responsible for maintenance of grass, flowers, trees, shrubs, and other plants on grounds. Is responsible for mowing grounds and trimming trees, flowers, and shrubs. Keeps grounds watered. Sprays plants with fertilizers or insecticides.

Horticultural Inspector *DOT Code 168.287*

Inspects horticultural goods to be sure they meet government standards. Inspects such goods as fruits, grains, vegetables, tobacco, and nursery items. Evaluates goods according to size, color, and condition. Issues certificate that identifies the quality or grade of a product. Rejects products that do not meet government standards. Consults with growers and informs them about planting, packing, and loading techniques. Also assists growers with the care and prevention of diseased plants.

Landscape Gardener *DOT Code 407.181*

Plans lawns. Maintains lawns and grounds for private individuals and industry. Is responsible for removing dead plants and planting new plants. Maintains general care of grounds which includes mowing, replanting, cultivating, trimming, spraying, and fertilizing.

Nursery Manager *DOT Code 406.168*

Is responsible for the management of the nursery. Has knowledge of soil content and life processes of plants, especially those processes involved with seed germination. Selects the seeds, trees, shrubs, flowers, and other plants to purchase and plant. Selects equipment and supplies needed for the nursery. Instructs customers in good growing techniques. Supervises other nursery workers.

Scientific Helper, Botanist *DOT Code 199.384*

Collects specimens for botanist. Assists botanist in conducting plant research. Prepares laboratory specimens for microscopic study by botanist. May conduct routine laboratory tests on plants.

Timber Cutter *DOT Code 940.884*
(Faller)

Fells trees. Is responsible for cutting tree at most advantageous angle. Selects angle so that tree does not harm people, other trees, or itself as it falls. Uses machinery (chainsaw).

Tree Surgeon *DOT Code 409.181*

Prunes trees and shrubs (usually in yards and parks). Sprays trees and uses other protective techniques to ward off attack by insects and disease. Cares for diseased trees. May plant trees.

Vegetable Grower *DOT Code 403.181*
(Garden Farmer)

Is responsible for overall operation of vegetable farming. Selects and purchases seeds for planting. Is responsible for cultivating and harvesting. May raise several vegetables or may specialize in the growing of a single vegetable (Onion Farmer, Potato Farmer). Supervises workers in vegetable growing.

Microbiology

Microbiology is that branch of the biological science family that studies microscopic organisms. Microbiologists study the growth, structure, and general characteristics of bacteria and other microorganisms. They are often employed in universities, medical and public health agencies.

Occupational Descriptions: Occupations Requiring College Training

Microbiologist *DOT Code 041.081*
(Bacteriologist)

Studies bacteria, viruses, molds, yeasts, protozoa, and other organisms of microscopic and sub-microscopic size. Prepares cultures of bacteria and other microorganisms. Identifies morphological and physiological characteristics of organisms. Conducts research to identify the effects of microorganisms on plant and animal tissues. Uses chemical analysis to investigate substances (alcohol, acid) produced by bacteria and other microorganisms. May specialize in a specific aspect of microbiology. Specific areas of specialization are listed below.

Dairy Bacteriologist: Studies bacteria and other microorganisms with respect to their influence on the processing and spoilage of milk.

Fishery Bacteriologist: Studies the harmful effects of bacteria and other microorganisms on fish.

Food Bacteriologist: Identifies the means by which bacteria and other microorganisms interfere with the spoilage and general deterioration of food (except milk).

Industrial Bacteriologist: Studies the beneficial and destructive effects of bacteria and other microorganisms in the manufacture of industrial fermentation products (beverage alcohols, glycerine, amino acids).

Medical Bacteriologist: Identifies and studies organisms that cause infectious diseases. After inoculating laboratory animals, tests the virulence of pathogenic bacteria. May specialize in a specific aspect such as the study of blood serum and blood groupings (Serologist). May study the methods by which the body fights off infection (Immunologist).

Pharmaceutical Bacteriologist: Conducts research to determine the most efficient methods for the manufacture of pharmaceutical products (penicillin, streptomycin, vitamins) that depend upon the actions of microorganisms for their formation.

Soil Bacteriologist: Studies the relationship between microorganisms in the soil. Studies soil fertility. Investigates conditions resulting from the effect of microorganisms on such soil processes as nitrogen fixation and ammonia production.

Microbiology Technologist *DOT Code 078.281*

Performs tests with human or animal body tissues to obtain evidence of disease or parasites. Often obtains tissue material from an autopsy. Cultivates, isolates, and identifies bacteria and other microorganisms. Performs tests according to instructions and requests of a pathologist.

Mycologist *DOT Code 041.181*

Studies poisonous, edible, and parasitic fungi. Performs research to obtain information about the life processes of fungi. Applies research findings to the medical and agricultural fields for use in the development of medicines, molds, and yeasts. Studies the anatomy, physiology, classification, and general growth of fungi.

Occupational Descriptions: Occupations Not Requiring College Training

Cleaner, Laboratory Equipment *DOT Code 381.887*

Cleans laboratory equipment such as beakers, metal containers, and machines. Mixes solutions used in the cleaning process. Keeps general laboratory area clean. May sterilize specific materials in an autoclave.

Scientific Helper *DOT Code 199.384*
(Laboratory Assistant)

Collects specimens and other microbiological materials for microbiologist. Assists microbiologist with research. May conduct routine laboratory tests under the direct supervision of a microbiologist.

Zoology

Zoology is the science that studies animal life. It is concerned with all aspects of animal life such as distribution, classification, life processes, and life histories for all types of animals.

Occupational Descriptions: Occupations Requiring College Training

Anatomist *DOT Code 041.081*

Examines the structure and form of animals. Dissects and observes animals as a means of comparing and contrasting one species with another. May dissect large organs or minute aspects of organs. Conducts research not only for the increased understanding of animals, but also for its application toward understanding disease in humans.

Animal Cytologist *DOT Code 041.081*

Studies animal cells. Uses microscope to study cell structure. Investigates the influence of chemical and physical factors on cell growth. Compares and contrasts normal cell with diseased or malignant cells.

Animal Ecologist *DOT Code 041.081*

Conducts research on the effects of the environment on the physical characteristics and life processes of animals. Performs work similar to that of zoologist with emphasis on environmental aspects.

Animal Husbandman *DOT Code 040.081*
(Animal Scientist)

Conducts research to obtain information on the selection, feeding, and breeding of animals (horses, swine, cattle, sheep, and pet animals). Studies breeding habits to develop better animal strains. Investigates management and marketing techniques. Studies housing, sanitation, and disease control. May specialize in feeding (Animal Nutritionist). May specialize in the care of a specific animal or herd (Dairy Husbandman).

Animal Paleontologist *DOT Code 024.081*

Studies fossilized remains of animals. Assembles and classifies animals by their family, age, and evolutionary development. Investigates plants in relation to their formation within the earth's crust in an effort to explain present, past, and future evolutionary development.

Animal Physiologist *DOT Code 041.081*

Studies the functioning of the life processes of animals. Conducts research to obtain better understanding of the functioning of the organs (heart, liver, lungs). Conducts research to ascertain the effects of the environment on animal life. Research results are often used by the medical profession. May specialize in the study of a specific organ.

Animal Taxonomist *DOT Code 041.081*

Identifies and classifies animals. Dissects animals to obtain more accurate information about animal anatomy. Conducts research to classify unknown species. May specialize in the study of a specific animal or class of animals.

Biochemist *DOT Code 041.081*

Studies the chemical processes of living organisms. Conducts chemical analyses to determine the action of drugs, foods, vitamins, enzymes, and allergens on living organisms and specific body functions. Studies the chemistry of such life processes as breathing and digestion. Conducts research on such body processes as antibody formation and blood corpuscle functioning. May specialize in the study of a specific substance or process and be designated by that title (Clinical Chemist, Enzyme Chemist, Steroid Chemist, Pharmaceutical Chemist).

Biophysicist *DOT Code 041.081*

Studies the physical principles of living cells and organisms. Conducts research into the physics responsible for such life processes as hearing and seeing. Studies the use of x-rays and their beneficial and detrimental effects on cells. Investigates the effect of atomic isotopes on cell transformation. May study a specific area such as radiology and its effect upon the treatment of cancer.

Biostatistician *DOT Code 020.188*

Conducts experiments to obtain statistical data for use in biological experiments. Studies the principles related to the life processes of animals. By using statistical techniques, the biostatistician compares the relationship between groups of organisms. Obtains statistical data relating to the cause and effect of a specific disease. Conducts research to determine the efficiency of a particular treatment plan. May specialize in the study of plants, animals, or both plants and animals.

Curator, Zoo *DOT Code 102.118*

Is responsible for the administrative affairs of the zoo. Obtains, develops, and supervises the collection of new specimens. Directs the exchange of specimens for loan. Conducts field trips. Conducts research. Supervises and coordinates the work of all personnel at the zoo.

Entomologist *DOT Code 041.081*

Studies insect life. Investigates the ways that insects affect other life. Identifies and classifies insects. Conducts research to develop insecticides that will aid in the control on harmful insects. Studies methods for effectively using beneficial insects.

Fish Culturist *DOT Code 041.168*

Is responsible for the supervision of workers engaged in cultivating fish for stocking lakes, streams, and ponds. Collects, studies, and interprets data on fish habits, growth, and migration. Also studies feeding habits. Collects information about the enemies of fish. Studies water conditions where various species of fish breed. Directs workers as they attempt to recreate satisfactory living conditions in ponds. Directs incubation of spawn. Supervises the treatment of spawn and fry. May specialize in the study of a specific fish.

Geneticist *DOT Code 041. 081*

Studies heredity. Specializes in the study of the mechanisms by which biological characteristics are transmitted from one generation to the next. Is responsible for performing experiments to determine the biological factors prominent in the origin, development, and transmission of inherited characteristics. Conducts research to improve animal breeds. Seeks methods for altering traits or producing new ones. May study both plants and animals or may specialize in the study of animals (Animal Geneticist or Animal Breeder) or plants (Plant Geneticist or Plant Breeder).

Histopathologist *DOT Code 041.181*
(Histologist)

Prepares and studies animal and human tissues to obtain information regarding disease. Develops stains for isolating tissues. Studies tissues by microscopic investigation to provide information on the cause and progress of pathological conditions. May conduct research to obtain data on body functioning. Supervises laboratory personnel in their preparation of tissue sections.

Parasitologist *DOT Code 041.081*

Investigates the characteristics and life habits of animal parasites. Studies the effects and manner by which they attack other animals, including human beings. Studies transmission from host to host and ways to combat parasites. Conducts research on new techniques for treating parasitic infection.

Pharmacologist *DOT Code 041.081*

Conducts experiments with animals to determine the effects of drugs and gases on tissues and organs of animals and human beings. Investigates methods for preventing and curing diseases. Conducts chemical analyses on food preservatives and colorings to determine if they do or do not cause toxic effects. Research is usually conducted with animals such as rats, guinea pigs, rabbits, etc.

Teacher, Zoology *DOT Code 091.228*
(Public School or College)

Instructs students in the study of zoology. Lectures and supervises laboratory sessions. May conduct research (especially at college level) to obtain more information on the life processes of animals.

Zoologist *DOT Code 041.081*

Studies all aspects of animal life including origin, life history, habitat, growth, development, and genetics. May conduct field trips to study animals in their natural habitat. Studies the effect of the environment on animal life. May specialize in the study of a specific animal.

Occupational Descriptions: Occupations Not Requiring College Training

Animal Breeder *DOT Code 419.181*

Breeds and cross-breeds animals to produce selective traits. Observes animals to be sure they remain in a healthy condition. Records animal data to be used in future breeding experiments. Keeps animal shelter clean. Feeds and waters animals. May specialize in the breeding of a specific animal (Dog Breeder).

Animal Caretaker *DOT Code 356.874*
(Animal Keeper)

Cares for animals (monkeys, rats, guinea pigs) used in medical and biological research. Keeps accurate records of animals' weight, diet, etc. Examines animals to be sure there are no signs of illness. Feeds animals and maintains records of type and amount of food eaten. Cleans (sometimes sterilizes) animal living quarters. May take animal from cage to research site. May assist in the collection of blood or urine samples.

Artificial-Breeding Technician *DOT Code 467.384*

Collects bull semen. Examines semen microscopically to check its motility and sperm count. Prepares semen for storage. May inseminate cows.

Dog Trainer *DOT Code 159.228*

Trains dogs to perform tricks, hunt, or race. May also train seeing eye dogs. Studies characteristics and breeds of dogs to ascertain specific traits. Also studies best methods for handling and training dogs.

Dog Warden *DOT Code 379.878*

Captures, impounds, and cares for stray dogs. Investigates complaints about uncontrolled dogs and rabid animals. Checks dog tags to be sure licenses are recent. Issues warnings to owners of uncontrolled or unlicensed dogs.

Farmer *DOT Code 421.181*

Raises various kinds of animals (also raises crops). Determines which livestock to raise. Is involved in breeding to obtain better strains. Feeds cattle. Treats diseased animals, often with help of veterinarian. Makes decisions regarding selling and slaughtering of animals. Is responsible for seeing that records are maintained on livestock. Keeps informed about livestock prices and decides when animals should be sold. May specialize in the raising of a specific animal (or crop) and be designated as dairy farmer, poultry farmer, game farmer, etc.

Fish and Game Warden *DOT Code 379.168*

Maintains biological data on wildlife in a specific area. Issues reports on availability and condition of wildlife. Checks to be sure that game laws are not being violated. In the case of a violation, makes an arrest and prepares information for court. May speak at schools and civic groups to inform them of the current game laws and availability of game.

Fish Farmer *DOT Code 436.181*
(Fisheries Aid)

Raises food fish for stocking fishing areas (lakes, streams, ponds). Strips eggs from female fish. Cares for eggs while they are developing into young fish. Feeds and cares for young fish in rearing ponds until they are ready to be transferred to larger ponds. May raise a specific fish, such as trout.

Fisher, Net *DOT Code 431.884*

Catches fish and other marine life. May work alone or with other members of crew. Fishers are often referred to by the gear they use such as line fisher, pot fisher, clam-dredge operator.

Gamekeeper *DOT Code 451.181*

Is responsible for the care and protection of animals on state game farms or private game preserves. Selects animals to be used for breeding stock. Selects eggs for use in artificial incubation. Prepares feed and water for animals. Monitors presence of predatory animals and sets traps for those that might harm or carry disease to animals on game farm.

General Helper, Laboratory *DOT Code 356.874*
(Animal Tender)

Cares for laboratory animals (guinea pigs, rabbits, monkeys, rats) used in scientific research. Checks animals to be sure they are not diseased. Is responsible for feeding, watering, and general care of animals. Is sometimes responsible for preparing food for special animal diets. May hold animals during injections and surgery. Keeps accurate record of animal weight, diet, and general health. Keeps animals' cages and living areas clean.

Kennelperson *DOT Code 356.874*

Cares for dogs in kennel. Feeds and waters dogs. Checks dogs to detect illness. Keeps kennels clean. Rebeds dog pens. Washes and grooms dogs. May be required to keep specific records on grooming, feeding, breeding, and general health of dogs. May perform these tasks with other domestic animals.

Kennel Master *DOT Code 349.368*

Is responsible for verifying credentials of all dogs entering dog races. Compares racing form information with actual dog credentials (verifies color, sex, owner, etc.). Checks to be sure dogs are placed in proper stall. Observes dogs to detect illness or injury. Reports any illness to veterinary personnel. Guards dogs to be sure they are not injured or disturbed by unauthorized persons. Cleans racing stalls after each race.

Medical and Health Careers

The medical and health careers include those areas of the medical health professions that provide a direct service to injured or diseased individuals or animals.

Occupational Descriptions: Occupations Requiring College Training

Cytotechnologist *DOT Code 078.281*

Is concerned with pathological condition of cells. Stains, mounts, and inspects slides (using microscope) to determine cellular damage. Reports findings to pathologist.

Dentist *DOT Code 072.108*

Diagnoses and treats disorders of the teeth and gums. Cleans, fills, and extracts teeth using dental instruments. Uses x-ray techniques to diagnose and advise patient about the extent of tooth decay or gum disorders. Instructs patients in proper dental care. Takes impressions for dentures. May specialize in a specific phase of dentistry such as extracting teeth (Exodontist), the correction of malpositioned teeth (Orthodontist), or the care of children's teeth (Pedodontist).

Dental Hygienist *DOT Code 078.368*

Is responsible for cleaning and polishing patients' teeth. Uses dental instruments to remove tartar deposits and stains from teeth and gums. Applies medication after the teeth have been cleaned. Charts problem areas (decay, poor gum tissue) for observation by dentist. May take and develop x-rays. Instructs patients and community groups in the care of teeth.

Inhalation Therapist *DOT Code 079.368*

Is responsible for administering oxygen and other gases to patients. Sets up and monitors equipment (oxygen tents, iron lungs, etc.). Checks patient to be sure no adverse effects are occurring from the use of the equipment. Instructs nurses and other attendants in the use of the equipment. Informs nurse of physician's prognosis and any special instructions given by physician. Sends out charge slips for services rendered.

Medical Technologist *DOT Code 078.381*

Performs laboratory tests to help in the diagnosis, treatment, and prevention of disease. Laboratory tests are performed on such specimens as blood, urine, pus, throat smears, etc. May take blood sample from patient. May work in specific area of laboratory performing tests relating to such specific areas as chemistry, bacteriology, hematology, blood bank, and serology. May assist with research involving new medicines.

Nurse *DOT Code 075.378*

Administers hospital care to patients directly or indirectly through the supervision of the head nurse. Observes patient recovery throughout an illness. Records patient's reaction to various treatments. Receives patient care instructions from the physician. Assists physician in examining the patient. May report patient's condition directly to physician or indirectly to the head nurse. May specialize in a specific duty (Surgical Nurse, Obstetrics Nurse).

Nurse, Registered: Performs duties described above. Training is longer than that required for a licensed practical nurse and duties require more specialized knowledge.

Nurse, Licensed Practical: Performs duties described above, as assigned by the registered nurse.

Optometrist *DOT Code 079.108*

Uses instrumentation to determine visual performance. Prescribes lenses and exercises to correct visual problems. Examines eyes to determine whether a disease condition exists. Refers patient to physician (Ophthalmologist) if diseased condition is identified or suspected.

Pharmacist *DOT Code 074.181*

Fills prescriptions as ordered by an authorized medical practitioner (usually physician or dentist). Is responsible for the weighing, measuring, and mixing of medicines. Keeps records of pharmaceutical supplies. Checks to be sure supplies are properly stored. May lecture civic groups on pharmaceutical matters.

Physical Therapist *DOT Code 079.378*

Treats patients with bone, muscular, nerve, and joint disorders according to directions prescribed by a physician. Uses massage, heat, water, and special equipment (ultraviolet lamps, ultrasonic machines) to relieve pain. Administers whirlpool baths. Instructs in the proper use of artificial appliances (braces, limbs, etc.).

Physician, M.D. *DOT Code 070.108*

Is concerned with the prevention, diagnosis, and treatment of illness. Examines patients using medical instruments to determine the extent of illness. May suggest that laboratory tests be performed to further identify the condition of the patient. Studies laboratory results in an effort to identify and prescribe treatment and drugs for the illness. Instructs patients in good health habits. A few of the medical specialties in which physicians may concentrate are listed below.

Anesthesiologist: Administers anesthetics to make patient insensitive to pain during surgery.

Cardiologist: Specializes in functioning of the heart and treatment of its disorders.

Gynecologist: Specializes in the functioning and treatment of the female generative organs.

Internist: Specializes in the functioning and treatment of the internal organs of the human body.

Obstetrician: Specializes in the treatment of women during pregnancy (includes prenatal, natal, and postnatal stages).

Ophthalmologist: Specializes in the treatment of eye diseases and injuries.

Pathologist: Studies tissues, fluids, and body secretions to determine the cause and progress of disease. Performs autopsies.

Pediatrician: Specializes in the medical care of children from birth through adolescence.

Psychiatrist: Specializes in the study of human behavior and mental illness. Treats behavioral and emotional disorders. May conduct therapy in individual or group sessions.

Radiologist: Uses x-rays and radioactive substances to diagnose and treat disease.

Surgeon: Performs surgery to treat injuries. May also perform surgery to improve patient functioning or to prevent disease.

Radiologic Technologist *DOT Code 078.368*

Treats patient by exposing diseased area of body to radioactive substances and x-rays as ordered by the physician. Is responsible for setting up, shooting, and developing the x-ray film. Maintains records of services offered and rendered. May specialize in taking x-rays of a specific body part.

Veterinarian *DOT Code 073.108*

Is concerned with the prevention, diagnosis, and treatment of animal disease. Conducts autopsies to determine the nature and cause of death. Prescribes medication for diseased animal. Performs surgery to remove or treat diseased organ. May inoculate entire herds against a particular disease (inoculate hogs for cholera). Instructs owners in the proper breeding and general care of animals. May specialize in the care of large or small animals. May specialize in a specific aspect of veterinary science (Veterinary Bacteriologist, Epidemiologist, Parasitologist, Pathologist, Pharmacologist, Physiologist, Radiologist, Virologist).

Occupational Descriptions: Occupations Not Requiring College Training

Dental Assistant *DOT Code 079.378*

Assists dentist. Seats patient in dental chair and readies instruments for dentist. Records information pertinent to patient's past and present dental history. Keeps dental instruments in working area clean. May assist in the mixing of dental amalgam. Supplies filling to dentist. May make appointments and collect fees.

Dental Ceramist *DOT Code 712.281*

Is responsible for constructing and repairing porcelain teeth according to the dentist's prescription. Is responsible for repeating the porcelain-application process until it meets the dentist's specifications.

Dental Laboratory Technician *DOT Code 712.381*

Is responsible for constructing and repairing full and partial dentures, crowns, inlays, and porcelain teeth. Uses electric grinders, electric furnaces, polishers, and teeth-color scales to produce the dental appliance prescribed by the dentist. Tests for occlusal harmony. May specialize in a specific appliance aspect (Crown and Bridge Technician, Metal Dental Technician, or Orthodontic Technician).

Denture Set-Up Person *DOT Code 712.381*

Uses wax models for mounting teeth. Uses the dentist's prescription for making dentures. Mounts teeth in prepared model. Removes excess wax from model. Uses electric tool to grind high areas on tooth.

Ear-Mold Laboratory Technician *DOT Code 777.381*

Is responsible for making plaster mold of ear that is used in hearing aids. Mixes and pours ingredients into a plastic mold. Grinds and polishes ear impression using electric grinder and polishing wheel.

Electrocardiograph Technician *DOT Code 078.368*

Is responsible for recording heart action by use of an electrocardiograph machine. Positions electrodes on specific body areas to record different types of heart action. Sends electrocardiograph tracing to physician for interpretation.

Electroencephalograph Technician *DOT Code 078.368*

Is responsible for recording brain impulses by use of an electroencephalograph machine. Fastens electrodes to different portions of the head. Observes patient throughout the recording of the electroencephalograph. Gives tracing to physician for analysis and interpretation.

First-Aid Attendant *DOT Code 354.878*

Aids physician in rendering medical aid at work sites (industry, mine, construction area, etc.). Renders first aid treatment to injured. Is responsible for administering oxygen and artificial respiration to sick or injured. Is responsible for changing beds and keeping equipment clean. May aid the physician in mending broken bones, applying antiseptic solutions. Works under direction of a physician.

Laboratory Assistant *DOT Code 381.887*
(Cleaner, Laboratory Equipment)

Is responsible for cleaning laboratory equipment. Mixes cleaning solutions according to directions given by chief laboratory personnel. Washes, rinses, and dries laboratory supplies. May label and fill laboratory supply bottles and deliver them to laboratory work areas.

Medical Assistant *DOT Code 079.368*

Assists the physician in the treatment of the patient. Interviews the patient. Takes temperature and pulse. May help the patient to become comfortable and ready for the examination. May hand the doctor instruments and supplies during the examination. Sterilizes instruments and keeps examination area clean. May collect fees and schedule appointments.

Midwife *DOT Code 354.878*

Provides direction and assistance to women during childbirth. Attends to the woman from the beginning of labor until a few hours after the child is born.

Nurse Aid *DOT Code 355.878*

Assists in the care of hospital patients. Bathes, dresses and feeds patients. Transports them to other areas of hospital. Changes linens and makes beds. Keeps patient's room and hospital facility clean. May clean and sterilize hospital supplies. Receives work assignment and supervision from the nurse or other professional member of the medical staff.

Orderly *DOT Code 355.878*

Is responsible for assisting with the medical care of patients. Receives work duties from nurse and other members of the medical staff. Bathes, shaves, and feeds patients. Takes temperatures and records pulse rates. Takes meal trays to patients. Takes patient to other areas within the hospital for x-rays, etc. Keeps room and hospital corridors clean. Makes hospital beds.

Pharmacy Helper *DOT Code 074.387*

Performs pharmacy tasks under the direction of a pharmacist. Keeps pharmacy work areas clean. May mix such preparations as mouthwash. Checks inventory and orders stock supplies. Washes and sterilizes pharmacy beakers and bottles. Labels pharmacy supplies. Delivers pharmacy orders to homes.

Psychiatric Aid *DOT Code 355.878*

Is responsible for caring for mentally ill patients. Receives work duties from nurse or other member of the medical staff. Helps patient become adjusted to hospital living. Helps patient dress and eat. Observes patient behavior and reports any abnormal behavior. May escort patient to library, recreational facilities, and other areas on the hospital grounds.

Scientific Helper *DOT Code 199.384*
(Laboratory Assistant)

Collects specimens and other medical materials for the medical staff. May conduct routine laboratory tests under the direct supervision of the laboratory supervisor.

Veterinary-Hospital Attendant *DOT Code 356.874*

Cares for animals during their stay in the veterinary hospital. Receives work orders from a veterinarian. Carries animal from living quarters to treatment room. Places animal on treatment table and applies restraints needed to keep animal on table. May administer anesthetic or other medications as directed by a veterinarian. Keeps instruments sterilized. Cleans the hospital area. Bathes and grooms animals. Prepares and feeds animal food.

Educational Requirements

Occupations Requiring College Training

Most of the occupations listed beneath the college training category require at least a bachelor's degree which usually requires four years of college training. Although some of the entry level positions in the biological science area are held by persons possessing bachelor's degrees, most of the higher level positions are awarded to persons possessing advanced degrees, and often doctorates (which usually require eight years of college training). More specific information about the amount of educational training required may be obtained from the *Handbook of Job Facts*.

Occupations Not Requiring College Training

In today's world, a high school diploma is usually required for most occupations. This trend seems to be true for the biological sciences area. Thus the student should plan to earn a high school diploma before seeking employment in any of the areas described. Also in most occupational areas, on-the-job training will be required. In other instances special training, usually of one year or less, may be required.

Interests and Aptitudes

The student who plans to pursue a career in the biological sciences area should show an interest in and aptitude for science. This may be evidenced in any number of ways. Usually the student has enjoyed his or her high school science program, and has performed well although grades should not be the only criteria. The student may have submitted projects for a science fair or he or she may have simply investigated scientific phenomena at home. Investigative work may range from a controlled experimental project to the keeping of tropical fish at home. Usually the student will have identified some area of biology as interesting to him or her and very often high school elective classes will have been selected in this area. Other students may have recognized a creative interest in scientific phenomena generally, but may not necessarily have geared this interest toward any scientific project or investigation.

In addition, students who plan to pursue a career in the biological sciences area should have an interest in living things. They should be curious and eager to obtain answers to their many questions about the living world. They should have perseverance and should not give up in the face of difficult situations. They should be conscientious and should want to see a job done accurately, carefully, and thoroughly. Objectivity, patience, sound judgment, honesty, ability to get along with others, and emotional stability are equally good traits for the future biologist to possess.

Places of Employment

Architect Firm
Business
College and University
Dental Clinic
Farm
Fish Hatchery
Florist
Game Preserve
Government
Hospital
Industry
Kennel
Medical Center
Museum
Nursery
Park
Pharmacy
Physician's Office
Public or Private School
Veterinary Clinic

Beneficial High School Classes

Although any number of classes would probably prove useful to the student planning to pursue a career in the biological sciences area, it is felt that in addition to biology some of the following classes should prove to be especially useful:

Agriculture (especially classes relating to the life sciences)
Anatomy
Botany
Chemistry
English Composition
Mathematics
Microbiology
Physiology
Physics
Zoology

Generally the student expecting to pursue a career in one of the life sciences (especially a career requiring college training) should elect to take as many science and mathematics classes as are feasible in his or her high school program. Upon completion of the first semester of college work, students will usually agree that one should not be concerned with "over-equipping" for college; instead the concern should be for "under-equipping." Thus students are strongly encouraged to talk with their high school counselors and science teachers and to plan to include appropriate science classes in their curriculums.

4

Careers Indirectly Related to the Biological Sciences

This chapter is designed to acquaint the reader with careers that are indirectly related to the biological sciences. That is, the specific career may not require extensive training in the biological sciences, but the work setting within a biological science area usually requires that the worker be familiar with specific aspects of the biological sciences. For example, a person pursuing a career as a medical librarian probably would not have extensive training in the biological sciences. On the other hand, in order for that person to perform his or her occupational role, he or she must be familiar with the biological sciences, especially those aspects relating to medicine.

The arrangement and description of occupational specialities within this chapter is similar to the format of the *Dictionary of Occupational Titles* and *Chronicle Occupational Briefs*. Each occupational listing includes a description of the occupation, worker task, work setting, and *Dictionary of Occupational Titles*' code. Occupational descriptions are separated into two categories based upon whether the occupation does or does not require college training. In some instances there was no definite answer as to whether the occupation should be listed as directly or indirectly related to the biological sciences or whether the occupation should be listed as requiring or not requiring college training. The final decision to place the occupation within a specific category was based upon the training requirements that were most often given for that occupation.

Additional information on each of the occupations may be obtained from the *Dictionary of Occupational Titles, Chronicle Occupational Briefs*, and the *Handbook of Job Facts*. The reader who is unfamiliar with the six-digit code of the *Dictionary of Occupational Titles* should refer to Volume II of the dictionary entitled "Occupational Group Arrangement of Titles and Codes." Additional assistance may be obtained from a guidance counselor.

Occupational Descriptions: Occupations Requiring College Training

Agricultural Engineer *DOT Code 013.081*

Applies principles from engineering and the biological sciences to agricultural problems. Is concerned with such problems as electric power, farm machinery, farm equipment, soil and water conservation. Designs equipment for measuring the effect of temperature and other weather conditions on plants and animals. Designs equipment to provide more efficient insect control. Applies engineering principles to design better methods of fertilization, crop rotation, crop storage, and irrigation. May supervise the installation of equipment and instruct in its use.

Audiologist *DOT Code 079.108*

Specializes in hearing disabilities. Uses speech audiometers and other instrumentation to determine patient's hearing efficiency. Plans auditory training and other rehabilitative services. May conduct research on audiology problems and design audiology equipment. May specialize in such specific areas as pediatric audiology or geriatric audiology.

Audiometrist *DOT Code 078.368*

Administers audiometric (hearing) tests. Performs audiometric work under the direct supervision of an audiologist or physician. Is responsible for placing the earphones in correct position. Instructs the patient on how to perform the audiometric test. Sends the results of the audiometric test to an audiologist for interpretation and recommendations.

Coroner *DOT Code 168.168*

Investigates deaths thought to have occurred from other than natural causes. Has autopsies performed to determine the cause of death. Submits a written report of the findings. Testifies at court hearings. Is usually asked to investigate accidental, violent or unexplained deaths. Some states require the coroner to have medical or legal training.

County Agricultural Agent *DOT Code 096.128*

Instructs farmers on the latest farm techniques. Advises farmers about the best seeds to plant and best methods for crop rotation. Also aids farmers in solving such farm problems as those relating to disease, soil, and harvesting. Is responsible for maintaining accurate agricultural records. Is active in helping promote extension programs. Gives lectures to keep the community informed on available services.

Curator, Medical Museum *DOT Code 102.118*

Is responsible for organizing medical collections at the museum. Maintains constant contact with the medical profession, other scientists, and organizations to exchange scientific data. Conducts research to ascertain pertinent information on new (and old) specimens acquired by the museum. Makes research information available. May publish research findings. May organize trips to other museums. Is responsible for the total administration of the museum and its employees.

Dietitian *DOT Code 077.168*

Directs the food service operation. Is usually employed in schools, hospitals, restaurants, hotels, government, and industrial cafeterias. Is responsible for planning menus, selecting and purchasing equipment, food, and supplies. Directs the work of others in the dietetic service area. May teach others about good diets. May conduct research.

Dietitian, Therapeutic *DOT Code 077.128*

Supervises the preparation of therapeutic diets as prescribed by a physician. Confers with the medical staff and helps plan a special diet for patients (e.g. special diet for a diabetic). Checks prepared meals to be sure they conform to the specifications set forth in the therapeutic diet. Instructs patients and their families about the correct foods to prepare and techniques for preparing them.

Editor, Scientific Book *DOT Code 132.068*

Reads book manuscripts. May suggest changes in the manuscript. Is responsible for checking for accuracy, clarity, and general suitability for publication. Confers with author regarding changes in the manuscript, royalties, publication dates, and copies.

Food and Drug Inspector *DOT Code 168.287*

Inspects establishments which manufacture and sell foods and drugs to be sure that legal standards of purity and sanitation are enforced. Informs proprietors of findings and suggests improvements needed to meet standards. Takes legal action to force the closing of any establishment not complying with standards. Subjects products to specific chemical tests to be sure standards are met. Destroys products that do not meet standards. May inspect a number of products or may inspect a specific product (Fish Inspector, Drug Inspector).

Forest Engineer *DOT Code 005.187*

Is responsible for supervising engineering activities concerned with the removal of logs from the timber area. Designs transportation systems (roads or rail) for removing logs. Draws maps to illustrate the topographical features of the land where the desired log supply is located. Designs better methods (bridges, culverts, docks) for removing logs from the area.

Hearing Clinician *DOT Code 079.108*

Administers audiology tests to determine hearing threshold. Works under the direct supervision of an audiologist. May direct hearing determinations within the school setting. Informs patient, family, or teacher of hearing disorders and methods for correcting the disorders.

Home Economist *DOT Code 096.128*
(County Home-Demonstration Agent)

Instructs families and groups on nutrition, child care, home furnishing, and clothing. May visit homes to teach families proper techniques in any of the above areas.

Hospital Administrator *DOT Code 187.118*
(Superintendent, Hospital)

Directs and coordinates activities of hospital and hospital personnel (administrative, medical and nursing staff). Works under the direction of a hospital governing board. Keeps community informed of hospital services and activities. Is responsible for recommending hospital improvements (building and equipment) to the board.

Landscape Architect *DOT Code 019.081*

Plans and prepares site areas (grounds and landscaping) for such projects as subdivisions, parks, resort areas, airports, schools, hospitals, museums, and recreational areas. Obtains data on geographic conditions such as soil, water, and drainage. Discusses plans with client, architect, engineers, and others assigned to project. Confers with construction personnel to be sure landscape specifications are met.

Medical Engineer *DOT Code 002.081*
(Biomedical Engineer)

Develops and designs medical equipment. Uses engineering knowledge and applies fundamental engineering principles to the medical fields. May sketch, draw, or analyze medical equipment. May conduct research and develop the design of more efficient equipment. Usually works in cooperation with physicists and others in the medical field to design machines (e.g. heart-lung machine).

Medical Librarian *DOT Code 100.168*

Is responsible for the management of the library or that portion of the library concerned with medical information. Searches literature and compiles lists of materials available in specific medical subject areas. Arranges for the housing of special collections and journals. Keeps current list of special equipment available (film strips, motion pictures).

Medical-Record Librarian *DOT Code 100.168*

Is responsible for maintaining records of hospital and clinic patients. Records physician's diagnosis, treatment, and other medical information in record.

Medical Social Worker *DOT Code 195.108*
(Caseworker)

Counsels and aids patients and their families in overcoming or adjusting to personal, social, or emotional problems that interfere with their medical well-being. Interprets patient's personal, social, or emotional problems to the medical team. Counsels patients to help them understand their problems and the need for accepting and following the medical care suggested. Also may counsel the patient's family.

Oceanographer *DOT Code 024.081*
(Geologist)

Often referred to as a marine geologist. Studies the structure of the ocean bottom. Is interested in such features as rocks, surface markings, and general structure of ocean bottom. Conducts investigations to study plant and animal life in the ocean. Also studies chemical composition. May specialize in such specific aspects of the ocean as biological, physical, or chemical conditions.

Occupational Therapist *DOT Code 079.128*

Directs the occupational training program for patients in a hospital or other medical setting. Plans patients' schedules to include occupational training that may help them to become more employable after leaving the hospital, or to become better adjusted physically and mentally during their hospitalization. Instructs patients in manual arts, creative arts, games, and crafts.

Physical Anthropologist *DOT Code 055.088*

Studies past and present human populations to ascertain their growth patterns, physical characteristics, heredity, and general human form. Studies and measures physical characteristics of existing populations. Also studies characteristics of human fossils.

Psychiatric Social Worker *DOT Code 195.108*

Counsels patients and their families to help them reduce emotional disturbances that precipitate more severe psychiatric problems. Works as a member of the medical team which is supervised by a psychiatrist. Keeps the medical team informed of family, social, and emotional factors that may contribute to the patient's illness.

Psychologist *DOT Code 045.108*

Studies human behavior. Diagnoses problem areas (behavioral, emotional, perceptual, learning, etc.) and plans treatment to rehabilitate the patient. Observes the patient directly in interviews and therapy sessions. Administers tests to collect additional data about the client. May consult with the psychiatrist or other members of the psychological team. May teach and conduct research. May specialize in a specific approach or work setting (Clinical Psychologist, Counseling Psychologist, Engineering Psychologist,

School Psychologist, Industrial Psychologist, Experimental Psychologist).

Public Health Service Officer *DOT Code 187.118*

Is responsible for the administration of a city or county public health program. Inspects hospitals and other institutions to be sure that public health standards are met. Gives lectures to community groups to help them become aware of public health standards. Helps to establish free health care centers (tuberculosis, cancer). Prepares reports concerning institutions inspected.

Rating Specialist, Occupational *DOT Code 169.168*

Evaluates the veteran's potential for employment. Advises the disability claims board of the degree of disability present and the claim that should be awarded. Collects data about the veteran's previous work and present degree of disability. Evaluates the veteran's work potential and determines whether the veteran can or can not meet the physical, educational, and mental demands of the job.

Salesperson, Chemicals and Drugs *DOT Code 266.258*

Sells chemicals or pharmaceuticals to hospitals, medical suppliers, universities, industries, and agricultural firms. Needs knowledge of product and its use within the firm. May need to demonstrate the product. May need to compare its usefulness with another product. Other products which require similar sales positions include cattle and poultry food supplements, dental and medical equipment and supplies, farm and garden equipment and supplies, mortician supplies, veterinary supplies, pet supplies, and surgical appliances.

Speech and Hearing Clinician *DOT Code 079.108*

Offers remedial services to persons having speech and hearing disorders. Administers audiological tests to determine the extent of hearing impairment. Instructs the patient in proper articulation and voice control.

Speech Pathologist *DOT Code 079.108*

Diagnoses and treats speech and language problems. Is responsible for directing the program to treat the disorder. Conducts research to ascertain the cause and best treatment for a particular disorder. Counsels patient and family about the disorder.

Occupational Descriptions: Occupations Not Requiring College Training

Animal Eviscerator *DOT Code 525.887*

Eviscerates (removes organs from) animals, usually at a meat packing plant or slaughterhouse. Bisects animal carcass. Removes organs from carcass. Cuts head from carcass. May wash carcass with hose.

Animal Sticker *DOT Code 525.884*

Prepares animal for butchering by severing the jugular vein. May remove head and skin after blood has drained. May serve as animal sticker for a specific type of animal such as hog, sheep, or calf.

Biological Photographer *DOT Code 143.382*

Uses photographic techniques to illustrate medical and biological phenomena. Uses time-lapse and ultraspeed photography to produce more definitive pictures. Provides photographic illustrations for scientific journals and teaching sessions. Makes lantern slides and transparencies. May conduct research to ascertain better methods for presenting biological materials. May design equipment.

Optician *DOT Code 713.381*

Is responsible for operating machines that make lenses for eye glasses. Makes and assembles lenses according to prescription specifications. Mounts lenses in frames.

Photographer, Scientific *DOT Code 143.282*

Photographs plant and animal tissues. Develops pictures. Takes pictures to illustrate scientific phenomena. May do photography through the aid of a photomicroscope.

Taxidermist *DOT Code 199.281*

Mounts skins of birds and animals. Tries to preserve animal in natural state. Uses wire and papier mâché molds to mount the specimens. Attempts to give accurate presentation of skin, feathers, hair, muscles, eyes, teeth, claws, and other features of the animal.

5

Implementation

Although much information is available on careers in the biological sciences, one of the most vexing problems confronting teachers and counselors is its implementation. Among the prevalent myths is the notion that students will obtain information about careers on their own. While some students will seek career information independently, considerable evidence is available (Hansen, 1972) to indicate students do not obtain the information they need and desire. In describing this situation, Hansen writes:

> Students do not get the help they need or want in exploring themselves and careers and the help they get is based on outmoded matching models of vocational guidance that do not adequately take into account knowledge about the changing nature of the individual and his values, the changing world of technology and work, the lifelong process of career development and career decisions, and the psychological meanings of work and leisure to different groups and individuals [Hansen, 1972].

It is the author's belief that the classroom teacher should be responsible for introducing students to careers in the discipline that he or she is teaching, and career information should include careers that require non-college as well as college training. The author is not suggesting that career education is the complete responsibility of the classroom teacher. In fact, it is simply not possible for the teacher to do the entire job. Certainly the counselor and other school personnel must be involved. Yet it is believed that the teacher is often more knowledgeable about the various careers available in his or her discipline than any other individual in the school. Also the design and implementation of career education for a number of students in a group setting would probably be more appropriately and advantageously executed in the classroom. Thus the beginning phases of career education might occur in the classroom. Then, as specific questions requiring special assistance arise, the student(s) should be referred to the counselor. Counselors are usually in a position to assist students, either individually or in groups, with their long-range career plans. Counselors are familiar with employment trends and training, and they are adept at comparing external job requirements with such internal characteristics as values, attitudes, and aptitudes.

Although several methods for implementation are suggested in this chapter, there is no single, "best" method. Instead, the reader is encouraged to use some of the methods and then adapt them to his or her work setting. While some of the methods might be classed innovative, others have been in practice for several years. In addition, it should be stressed that although the author has attempted to include many of the prominent innovations, the list is not inclusive. Hopefully, this chapter will stimulate reflective thought that will result in more innovative practices in the schools.

Audio-Visual Materials

Bulletin Board Displays

A bulletin board display with emphasis on a specific biological science career can help the student widen his or her understandings of a particular career. In many instances it can help students consider an entirely new career. The bulletin board might be initiated as a student project to be integrated with a career class. Care should be taken to be sure that a new bulletin board is prepared regularly and frequently (weekly).

Cassettes or Tapes

Many cassettes or tapes have been developed by educational firms. Some of these materials may present workers describing their careers. Others present more formal interviews with workers.

A class might elect to produce their own cassette or tape. Before the students undertake this project, an outline should be prepared. This outline would include the information that the student desires to obtain as well as the procedure for obtaining the information. Students might want to role play the various occupations presented on the tape.

Filmstrips and Slides

The visual presentation provided by a filmstrip or slide will often hold the student's attention and thus create a better understanding of the occupation. Several educational firms have filmstrips and slides available for purchase or rental. Some professional organizations will lend them free. In addition, the reader is encouraged to peruse the *Filmstrip Guide* and *Educators' Guide to Free Slidefilms.* Both guides are published annually; the former has monthly supplements. Slides can be made locally at a relatively small cost to the school. Members of the school's audio-visual department will usually help. Also school alumni often enjoy this type of project.

Motion Pictures

One of the most useful of the audio-visual materials is the motion picture. It can bring the worker and the work setting into realistic dimensions for the student. Many companies are involved in the production of career education films. The reader is cautioned to be sure that the film is up-to-date. Many occupations change rapidly, and the film user should be sure that the content is relevant and timely. In addition, the reader should peruse the *Educational Film Guide.* The guide is published annually with monthly supplements. Usually it is more practical to rent films. They are expensive to purchase, and the content can become obsolete rapidly.

Radio and Television

Some radio and television stations offer programs on career education. The teacher should check program listings at educational stations as well as local stations. In addition, the school's career projects might be described by local stations.

Personal Visitations

Business and Industry Tours

Students can obtain a wealth of information about the world of work when they visit a worker on the job. Usually some members of the school faculty accompany the students. Not only does this give the student more information about careers, but it also allows the faculty to obtain more information about specific careers within the community.

Career Classes

Many high schools can assist students with career exploration by offering a class that is specifically directed toward the study of careers. This class should be basic, pursuing such fundamental questions as what does a biologist do, how does one enter a biology career, etc. Interests, needs, values, aptitudes, economic factors, job behavior, training requirements, and occupational trends should be discussed. In some instances specific tests measuring interest, maturity, and aptitude would be useful. In other instances, guest speakers and visitations might be integrated within career units.

Career Conference . . . Career Day

Career conferences and career days are similar. The major difference is that the conference usually allows for a two- or three-day session instead of the one day typically provided by the career day format. Career conferences and days provide good opportunities for students to hear representatives of specific occupations. In many instances the speakers find it useful to supplement their speeches with films and/or filmstrips. It is often beneficial to encourage students to select and invite the speakers. This gives students more involvement and usually stimulates more interest in the program.

College and School Tours

Students should be encouraged to visit the college and technical school campus. These tours offer opportunities for the student to compare schools as well as curricula. In addition, the student may want to schedule an interview with a member of the admissions staff to obtain additional information about admissions criteria.

College Day and/or Night

College day or night programs offer additional opportunities for the student to obtain information about college programs. In addition, this type of program affords the opportunity for parents to talk with college representatives about the various college programs, expenses, regulations, etc.

Interviews

Students can gain a wealth of information about careers from an interview. Each student could select a unit of biology that is of particular interest to him or her. Then the student could arrange an interview with an individual whose career requires training in that specific area of biology. This technique could be more interesting if the student taped the interview and shared the results with other members of the class. In a very short time, a useful set of career tapes would be available. These tapes could be integrated into the specific unit of biology study.

Speakers

Individuals representing various biological science careers could be invited to the classroom to give students a better understanding of their work activities. They should be encouraged to freely discuss the unsatisfactory as well as the satisfactory aspects. In addition, the students might prepare an outline listing the aspects of the career that they would like the speaker to discuss. A discussion with the speaker by a student committee would be beneficial before the presentation is made.

Student Work Programs

Many schools combine classroom instruction with employment experience. The student receives basic classroom instruction at school and basic job skill training in an actual employment setting. Usually the student attends school half of the day and works the other half. While his or her on-the-job training is given by the employer, a member of the school staff visits the work setting and serves as a liaison person between the employer and school. Usually the student receives academic credit for this work and in some instances a small salary.

This program affords the opportunity for a student to decide realistically whether he or she wants to pursue this career option. It also allows the student to obtain work experience and career information that should prove to be an asset as he or she seeks future education and employment. Other similar programs are provided under such titles as on-the-job training programs, apprenticeships, Armed Forces training, Manpower Development and Training Act programs, Job Corps, and Neighborhood Youth Corps.

Work Demonstration

It is often advantageous for students to observe workers engaged in typical job activities. If time permits, it would be even more advantageous for the student to see different workers on the job. If time can be arranged, the student should plan to spend an entire day with the worker.

Publications

Books, Newspapers, Journals

The school newspaper and local newspapers offer opportunities for feature stories about careers in the biological sciences. The school library could offer a display of books and other materials relating to careers. The librarian could prepare a bibliography of career materials. In addition, several commercial sources prepare career publications. Among the many sources are Chronicle Guidance, Science Research Associates, and Careers, Inc. Also, the *Dictionary of Occupational Titles, Occupational Outlook Handbook,* and the *Handbook of Job Facts* present information about job settings, worker traits, worker tasks, employment outlook, places to seek employment, and salaries. Many publications are available from professional organizations and companies. The school librarian and counselor are good resources for locating and advising on the use of career materials.

Microfilm

The use of microfilm in reader-scanners and reader-printers is becoming increasingly prevalent. Various commercial sources prepare career information in this form. The VIEW system, originally a Title III Project, is available through state departments of vocational information. The advantages this system has over books are two-fold. First, there are two cards on each career, one showing national trends and needs, and the other giving local and state job descriptions and information. Second, the reader-printer can make an 8½″ × 11″ copy of any material which the student may then keep for his or her personal reference.

Resource People

Alumni

Graduates of the school could be invited to return to inform students about their careers in science. Alumni have usually had career experiences in many different areas that will interest students. They should certainly be encouraged to relate both satisfactory and unsatisfactory experiences. A list of available alumni and their respective careers should be made available for student referral.

Career Resource Center Personnel

Some states have career resource centers. In most instances, personnel from these centers will visit schools and help teachers plan and implement career units. The center usually has career materials (books, films, etc.) for loan to teachers and counselors. In some states mobile units are available for school visitations. Usually the career resource staff is available to conduct career workshops.

Community Resource People

This category offers a broad variety of people who can give insight into career opportunities within the community. Business and industry personnel are likely to be very well informed on careers within their firms. Sometimes these people will invite individual students to their work settings, or they may prefer to visit the school. This not only provides the student with additional career information but it allows the school staff to get to know these people. Members of professional organizations and service clubs are often willing to assist students and school personnel with career endeavors. Additional community resources include churches, county extension offices, and employment, government, and social agencies.

Counselors

Counselors are available to assist students with vocational planning. They have career materials that will stimulate the student as he or she thinks about his or her vocational choice. They are also helpful to the student who needs to explore careers in a number of different areas. Counselors are knowledgeable about training programs, worker traits, employment trends, work settings, and other aspects of work. In addition, counselors have career materials on hand that describe various careers. They are available to assist students as they peruse these materials. Counselors are good resource people for assisting teachers with the planning of career education units in the classroom. They are also good resource people for providing the names of companies and/or publishers that have career education materials.

Parents

Parents usually represent a wide variety of occupations. Many parents would welcome an invitation by a student or teacher to visit the classroom to discuss their careers. Some parents may prefer to speak to the students in a group setting and others may prefer to serve as members of a panel. Still others may prefer to talk with students individually or at their work setting. It is very helpful to keep a directory of parents who are willing to serve as speakers.

School Staff

Teachers, counselors, administrators, and other school personnel can often give valuable critiques of career opportunities within the community. Although the school faculty members may not be employed outside the school setting, they are often knowledgeable about careers held by relatives, friends, or others in the community. Many staff members will have obtained first hand information from summer employment. In addition, they are usually knowledgeable about college programs, especially those offered by the colleges they attended.

State Department of Education

State departments of public instruction have employees who are available to assist teachers, counselors, and other school personnel with career education. Contacts should be made with both the specialist in the specific subject area (e.g. biological sciences) and with the director of pupil personnel services. An individual in the pupil personnel services area (e.g. counselor, occupational specialist) usually has responsibility for career education.

Role Models

Career Clubs

Science clubs offer an opportunity for students to obtain additional information about careers. Guest speakers, audio-visual materials, and plant tours usually make good presentations at the meetings.

Games

Career games have been developed by educational firms. Also, some career games have been reported in journals (Boocock, 1967 and Varrenhorst, 1968). Boocock reports a career life game that can be played by any number of teams ranging from two to four players. Each team works with a case history. Rounds or decision periods are played, each of which represents one year in the life of the person depicted by the case history. The players allot time for the person to engage in activities in such areas as family, work, study, and leisure. The player's goal is to select those activities that will best allow the person to enjoy a satisfying life. Scores are computed for four areas: education, career, family, and leisure. The team with the highest score (usually at the end of 12 rounds) wins. Spinners and dice are used to build a chance or "luck factor" into the game. In playing the game, students acquire skills in decision making and completing college and job applications. The author states that the game may be played by anyone at the sixth grade level or higher. This game could be integrated into the career class. It could also provide an excellent setting for role playing.

Another type of career game could easily and inexpensively be prepared by the students. Several occupations could be listed on index cards. The student could select the three occupations that he or she likes best and the three he or she likes least. The teacher and students could then have the students discuss the factors that caused them to be interested or uninterested in the occupations. Such aspects as training, worker traits, status, people contact, and work setting should be considered.

Role Playing

Role playing affords students the opportunity to portray different types of careers. It also allows students to observe and react to the roles played by others. Fundamental to good role playing is obtaining information about the career before the role is assumed.

Summary

In summary, this monograph strongly advocates the implementation of career education in the high schools. Chapter 1 presents findings from several studies that suggest that students do need assistance with their career planning. Realizing that there is no correct way to present career education, the author has devoted the remaining chapters to presenting concrete materials and suggestions for implementing career education within the classroom setting.

All chapters within the monograph are basic in their approach, assuming that the student is being introduced to careers in the biological sciences. Since this is an assumption, the teacher should assess the "entering level" of the student. A key problem for the teacher is to determine where the student is in his or her career development, and to be sure that the career information does not turn out to be "old stuff." Only when the entering level of the student has been determined can this monograph, or any monograph, be used appropriately and successfully.

APPENDIX

Professional Agencies and Organizations Supplying Materials and Career Information

Aerospace Medical Association
Washington National Airport
Washington, D.C. 20001

American Association of Colleges of Pharmacy
Office of Student Affairs
8121 Georgia Avenue
Suite 800
Silver Spring, Maryland 20910

American Association of Dental Schools
1625 Massachusetts Avenue, N.W.
Washington, D.C. 20036

American College of Hospital Administration
840 North Lake Shore Drive
Chicago, Illinois 60611

American Dental Assistants Association
1625 Massachusetts Avenue, N.W.
Washington, D.C. 20036

American Dental Association
Division of Career Guidance
Council on Dental Education
211 East Chicago Avenue
Chicago, Illinois 60611

American Dental Hygienists' Association
211 East Chicago Avenue
Chicago, Illinois 60611

American Dietetic Association
620 North Michigan Avenue
Chicago, Illinois 60611

American Library Association
50 East Huron Street
Chicago, Illinois 60611

American Medical Association
535 North Dearborn Street
Chicago, Illinois 60610

American Medical Record Association
875 North Michigan Avenue
Suite 1850
Chicago, Illinois 60611

American Nurses Association
2420 Pershing Road
Kansas City, Missouri 64108

American Occupational Therapy Association
6000 Executive Boulevard
Suite 200
Rockville, Maryland 20852

American Optometric Association
7000 Chippewa Street
St. Louis, Missouri 63119

American Pharmaceutical Association
2215 Constitution Avenue, N.W.
Washington, D.C. 20036

American Physiological Society
9650 Rockville Avenue
Bethesda, Maryland 20014

American Podiatry Association
20 Chevy Chase Circle, N.W.
Washington, D.C. 20015

American Psychiatric Association
1700 18th Street, N.W.
Washington, D.C. 20009

American Psychological Association
1200 17th Street, N.W.
Washington, D.C. 20036

American Society for Microbiology
1916 I Street, N.W.
Washington, D.C. 20006

American Society of Agricultural Engineering
2950 Niles Road
St. Joseph, Michigan 49085

American Society of Agronomy, Inc.
677 South Segoe Road
Madison, Wisconsin 53711

American Speech and Hearing Association
9030 Old Georgetown Road
Washington, D.C. 20014

American Society of Landscape Architects
1750 Old Meadow Road
McLean, Virginia 22101

Blood Services
P.O. Box 1030
Scottsdale, Arizona 85252

Marine Technology Society
Careers in Oceanography
1730 M Street, N.W.
Washington, D.C. 20036

National Association for Practical Nurse Education and Service, Inc.
122 East 42nd Street
New York, New York 10017

National Association of Colleges of Mortuary Science, Inc.
229 Park Avenue South
New York, New York 10003

National Center for Information on Careers in Education
1607 New Hampshire Avenue, N.W.
Washington, D.C. 20009

National Park Service
Department of the Interior
Career Employment
Washington, D.C. 20240

National Science Teachers Association
1201 16th Street, N.W.
Washington, D.C. 20036

Occupational Outlook Service
U.S. Department of Labor
Science Careers
441 6th Street, N.W. Room 2734
Washington, D.C. 20212

Registry of Medical Technologists
P.O. Box 4872
Chicago, Illinois 60680

Scientific Manpower Commission
1776 Massachusetts Avenue, N.W.
Washington, D.C. 20036

Society of American Foresters
1010 16th Street, N.W.
Washington, D.C. 20036

Soil Conservation Service
U.S. Department of Agriculture
Washington, D.C. 20250

Soil Science of America, Inc.
677 South Segoe Road
Madison, Wisconsin 53711

U.S. Civil Service Commission
Job Information Center
Washington Area Office
Civil Service Careers
Washington, D.C. 20415

U.S. Department of Agriculture Research Service
Animal and Plant Health Inspection Service
14th and Independence Avenue, S.W.
Washington, D.C. 20250

U.S. Department of Agriculture
14th and Independence Avenue, S.W.
Washington, D.C. 20250

Veterinary Medical Association
600 South Michigan Avenue
Chicago, Illinois 60605

BIBLIOGRAPHY

Asimov, I. *A short history of biology*. Garden City, New York: The Natural History Press, 1964.

Boocock, S. The life career game. *Personnel and Guidance Journal*, 1967, 46, 328–334.

Caullery, M. *A history of biology*. New York: Walker and Company Publishing Company, 1966.

Centers, R. Motivational aspects of occupational stratification. *Journal of Social Psychology*, 1948, 28, 187–217.

Conyne, R. K. and Cochran, D. J. Academia and career development. *Personnel and Guidance Journal*, 1973, 52, 217–223.

Dallas, J. D. Career education in a junior high school. *Personnel and Guidance Journal*, 1973, 52, 193–195.

Educational Film Guide, Standard Catalog Series. New York: H. W. Wilson Company.

Educators' Guide to Free Slidefilms. Randolph, Wisconsin: Educators Progress Service.

Filmstrip Guide. New York: H. W. Wilson Company.

Gardner, E. J. *A history of life science*. Minneapolis: Burgess Publishing Company, 1960.

Hansen, L. S. A model for career development through curriculum. *Personnel and Guidance Journal*, 1972, 51, 243–250.

Hopke, W. E. *The encyclopedia of careers and vocational guidance*. New York: Doubleday and Company, Inc., 1967.

Hoppock, R. and Novick, B. The occupational information consultant: A new profession? *Personnel and Guidance Journal*, 1971, 49, 555–559.

Hoyt, K., Evans, R., Mackin, E., and Mangum, G. *Career education: What it is and how to do it*. Salt Lake City: Olympus Publishing Company, 1972.

Isaacson, L. E. *Career information in counseling and teaching*. Boston: Allyn and Bacon, Inc., 1966.

Lang, Carole J. *Handbook of job facts.* Chicago: Science Research Associates, Inc., 1972.

Lanham, U. *Origins of modern biology.* New York: Columbia University Press, 1968.

Lyman, Elizabeth L. Occupational differences in the value attached to work. *American Journal of Sociology,* 1955, 61, 138–144.

McGuire, W. G. A three-phased career planning process. *Journal of College Student Personnel,* 1973, 14, 175.

Morrill, W. H. and Forrest, D. J. Dimensions of counseling for career development. *Personnel and Guidance Journal,* 1970, 49, 299–305.

National Academy of Sciences. *The life sciences.* Washington, D.C., 1964.

Norris, W., Zeran, F. R., and Hatch, R. N. *The information service in guidance.* Chicago: Rand McNally Company, 1960.

Patterson, R. (Ed.) *Need a lift?* (23rd ed). Indianapolis: The American Legion Education and Scholarship Program, 1973.

Reardon, R. C. The counselor and career information services. *Journal of College Student Personnel,* 1973, 14, 495–500.

Rook, A. (Ed.) *The origins and growth of biology.* Baltimore: Penguin Books, 1964.

Singer, C. *A history of biology.* New York: Abelard-Schuman, Inc., 1931.

Sirks, M. J. *The evolution of biology.* New York: The Ronald Press Company, 1964.

Stevenson, J. B. *An introduction to career education.* Worthington, Ohio: Charles A. Jones Publishing Company, 1973.

Super, D. E. *The psychology of careers.* New York: Harper and Row, Publishers, 1957.

U.S. Department of Labor, Bureau of Labor Statistics. *Occupational Outlook Handbook.* Washington, D.C.: U.S. Government Printing Office, 1972.

U.S. Department of Labor. *Dictionary of Occupational Titles* (3rd ed.). Washington: U.S. Government Printing Office, 1965.

Varrenhorst, Barbara. Innovative tool for group counseling: Life Career Game. *School Counselor,* May 1968, 15, 357–362.

Wysong, H. E. *Career education program.* Boston: Houghton Mifflin Company, 1973.

INDEX